ADDITION AND SUBTRACTION MASTERY

WITH ANSWERS

**FOR KIDS
AGE 3-10
1ST GRADE
TO 4TH GRADE**

NAME:

_____

CLASS:

_____

Thank you for choosing this book. This book is carefully designed by great educators for grade 1 to Grade 4 kids to master addition and subtraction. We have mixed 20 types of activities and puzzles to ensure that the kids enjoy every page of the book.

TABLE OF CONTENTS

Basic Addition

Find the sum.

1.
$$\begin{array}{r} 3 \\ +\ 5 \\ \hline \end{array}$$

2.
$$\begin{array}{r} 7 \\ +\ 4 \\ \hline \end{array}$$

3.
$$\begin{array}{r} 4 \\ +\ 2 \\ \hline \end{array}$$

4.
$$\begin{array}{r} 6 \\ +\ 4 \\ \hline \end{array}$$

5.
$$\begin{array}{r} 1 \\ +\ 2 \\ \hline \end{array}$$

6.
$$\begin{array}{r} 6 \\ +\ 7 \\ \hline \end{array}$$

7.
$$\begin{array}{r} 1 \\ +\ 4 \\ \hline \end{array}$$

8.
$$\begin{array}{r} 9 \\ +\ 8 \\ \hline \end{array}$$

9.
$$\begin{array}{r} 8 \\ +\ 1 \\ \hline \end{array}$$

10.
$$\begin{array}{r} 9 \\ +\ 7 \\ \hline \end{array}$$

11.
$$\begin{array}{r} 7 \\ +\ 8 \\ \hline \end{array}$$

12.
$$\begin{array}{r} 1 \\ +\ 9 \\ \hline \end{array}$$

13.
$$\begin{array}{r} 7 \\ +\ 2 \\ \hline \end{array}$$

14.
$$\begin{array}{r} 5 \\ +\ 9 \\ \hline \end{array}$$

15.
$$\begin{array}{r} 8 \\ +\ 6 \\ \hline \end{array}$$

16.
$$\begin{array}{r} 0 \\ +\ 2 \\ \hline \end{array}$$

17.
$$\begin{array}{r} 6 \\ +\ 8 \\ \hline \end{array}$$

18.
$$\begin{array}{r} 3 \\ +\ 3 \\ \hline \end{array}$$

19.
$$\begin{array}{r} 2 \\ +\ 1 \\ \hline \end{array}$$

20.
$$\begin{array}{r} 9 \\ +\ 9 \\ \hline \end{array}$$

21.
$$\begin{array}{r} 5 \\ +\ 3 \\ \hline \end{array}$$

22.
$$\begin{array}{r} 8 \\ +\ 2 \\ \hline \end{array}$$

23.
$$\begin{array}{r} 8 \\ +\ 4 \\ \hline \end{array}$$

24.
$$\begin{array}{r} 3 \\ +\ 7 \\ \hline \end{array}$$

25.
$$\begin{array}{r} 4 \\ +\ 7 \\ \hline \end{array}$$

26.
$$\begin{array}{r} 2 \\ +\ 6 \\ \hline \end{array}$$

27.
$$\begin{array}{r} 6 \\ +\ 1 \\ \hline \end{array}$$

28.
$$\begin{array}{r} 5 \\ + 4 \\ \hline \end{array}$$

29.
$$\begin{array}{r} 5 \\ + 1 \\ \hline \end{array}$$

30.
$$\begin{array}{r} 7 \\ + 9 \\ \hline \end{array}$$

31.
$$\begin{array}{r} 6 \\ + 2 \\ \hline \end{array}$$

32.
$$\begin{array}{r} 2 \\ + 2 \\ \hline \end{array}$$

33.
$$\begin{array}{r} 6 \\ + 5 \\ \hline \end{array}$$

34.
$$\begin{array}{r} 1 \\ + 8 \\ \hline \end{array}$$

35.
$$\begin{array}{r} 2 \\ + 7 \\ \hline \end{array}$$

36.
$$\begin{array}{r} 4 \\ + 5 \\ \hline \end{array}$$

37.
$$\begin{array}{r} 2 \\ + 5 \\ \hline \end{array}$$

38.
$$\begin{array}{r} 0 \\ + 6 \\ \hline \end{array}$$

39.
$$\begin{array}{r} 7 \\ + 1 \\ \hline \end{array}$$

40.
$$\begin{array}{r} 4 \\ + 8 \\ \hline \end{array}$$

41.
$$\begin{array}{r} 5 \\ + 6 \\ \hline \end{array}$$

42.
$$\begin{array}{r} 4 \\ + 3 \\ \hline \end{array}$$

Make Sum

What number should be added to the first number to make the second number?

43. 4
    +
    5

44. 2
    +
    5

45. 2
    +
    4

46. 4
    +
    9

47. 5
    +
    5

48. 4
    +
    6

49. 2
    +
    2

50. 2
    +
    8

51. 1
    +
    3

52. 3
    +
    7

53. 4
    +
    8

54. 5
    +
    8

55.
$$
\begin{array}{r}
3 \\
+ \\
\hline
5
\end{array}
$$

56.
$$
\begin{array}{r}
7 \\
+ \\
\hline
8
\end{array}
$$

57.
$$
\begin{array}{r}
2 \\
+ \\
\hline
6
\end{array}
$$

58.
$$
\begin{array}{r}
8 \\
+ \\
\hline
9
\end{array}
$$

59.
$$
\begin{array}{r}
2 \\
+ \\
\hline
3
\end{array}
$$

60.
$$
\begin{array}{r}
1 \\
+ \\
\hline
2
\end{array}
$$

61.
$$
\begin{array}{r}
6 \\
+ \\
\hline
7
\end{array}
$$

62.
$$
\begin{array}{r}
5 \\
+ \\
\hline
6
\end{array}
$$

63.
$$
\begin{array}{r}
5 \\
+ \\
\hline
7
\end{array}
$$

64.
$$
\begin{array}{r}
3 \\
+ \\
\hline
4
\end{array}
$$

65.
$$
\begin{array}{r}
3 \\
+ \\
\hline
9
\end{array}
$$

66.
$$
\begin{array}{r}
4 \\
+ \\
\hline
4
\end{array}
$$

67.
$$
\begin{array}{r}
1 \\
+ \\
\hline
8
\end{array}
$$

68.
$$
\begin{array}{r}
1 \\
+ \\
\hline
1
\end{array}
$$

69.
$$
\begin{array}{r}
1 \\
+ \\
\hline
4
\end{array}
$$

70.
$$\begin{array}{r} 3 \\ + \\ \hline 3 \end{array}$$

71.
$$\begin{array}{r} 3 \\ + \\ \hline 8 \end{array}$$

72.
$$\begin{array}{r} 1 \\ + \\ \hline 6 \end{array}$$

73.
$$\begin{array}{r} 4 \\ + \\ \hline 7 \end{array}$$

74.
$$\begin{array}{r} 2 \\ + \\ \hline 7 \end{array}$$

75.
$$\begin{array}{r} 8 \\ + \\ \hline 8 \end{array}$$

76.
$$\begin{array}{r} 6 \\ + \\ \hline 6 \end{array}$$

77.
$$\begin{array}{r} 7 \\ + \\ \hline 9 \end{array}$$

78.
$$\begin{array}{r} 3 \\ + \\ \hline 6 \end{array}$$

79.
$$\begin{array}{r} 7 \\ + \\ \hline 7 \end{array}$$

80.
$$\begin{array}{r} 5 \\ + \\ \hline 9 \end{array}$$

81.
$$\begin{array}{r} 1 \\ + \\ \hline 7 \end{array}$$

82.
$$\begin{array}{r} 6 \\ + \\ \hline 9 \end{array}$$

83.
$$\begin{array}{r} 1 \\ + \\ \hline 5 \end{array}$$

84.
$$\begin{array}{r} 6 \\ + \\ \hline 8 \end{array}$$

Pictorial Addition

Add the pictures

85. 🎁🎁🎁🎁🎁 + 🎁🎁🎁 = [ ]

86. 🎁🎁🎁 + 🎁🎁🎁 = [ ]

87. ✿✿✿ + ✿✿ = [ ]

88. 🍁🍁🍁🍁🍁 + 🍁 = [ ]

89. ★ + ★★ = [ ]

90. 🍁🍁🍁🍁 + 🍁🍁🍁🍁🍁🍁 = [ ]

91. ❀❀❀❀❀❀ + ❀❀❀❀❀ = [ ]

92. ★★★★★★★★ + ★★★★★★★★ = [ ]

93. ✿✿✿ + ✿✿✿✿✿✿✿✿ = [ ]

94. ✿✿✿✿✿✿✿✿ + ✿✿ = [ ]

95. ♥♥♥♥♥♥♥♥ + ♥ = [ ]

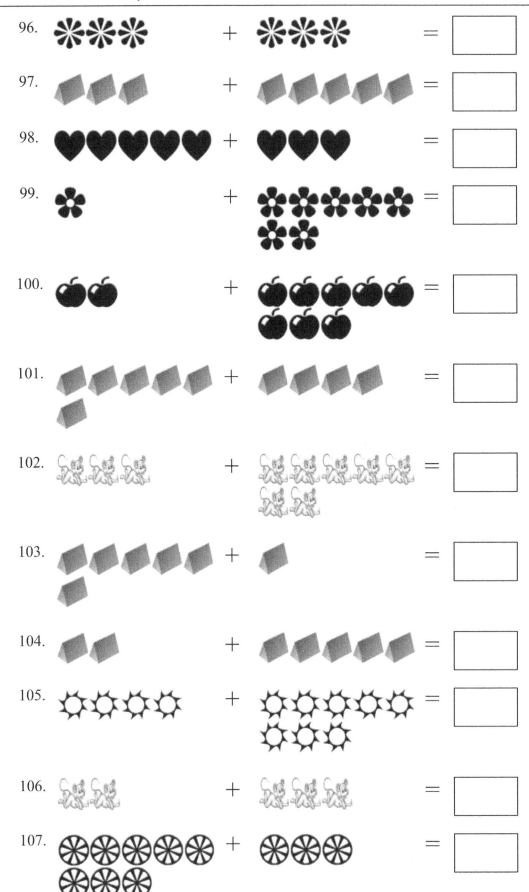

Basic Subtraction

Find the difference.

108.
$$\begin{array}{r} 2 \\ -\ 1 \\ \hline \\ \hline \end{array}$$

109.
$$\begin{array}{r} 15 \\ -\ 1 \\ \hline \\ \hline \end{array}$$

110.
$$\begin{array}{r} 5 \\ -\ 4 \\ \hline \\ \hline \end{array}$$

111.
$$\begin{array}{r} 11 \\ -\ 2 \\ \hline \\ \hline \end{array}$$

112.
$$\begin{array}{r} 13 \\ -\ 7 \\ \hline \\ \hline \end{array}$$

113.
$$\begin{array}{r} 2 \\ -\ 2 \\ \hline \\ \hline \end{array}$$

114.
$$\begin{array}{r} 17 \\ -\ 8 \\ \hline \\ \hline \end{array}$$

115.
$$\begin{array}{r} 4 \\ -\ 2 \\ \hline \\ \hline \end{array}$$

116.
$$\begin{array}{r} 10 \\ -\ 5 \\ \hline \\ \hline \end{array}$$

117.
$$\begin{array}{r} 18 \\ -\ 1 \\ \hline \\ \hline \end{array}$$

118.
$$\begin{array}{r} 5 \\ -\ 2 \\ \hline \\ \hline \end{array}$$

119.
$$\begin{array}{r} 7 \\ -\ 1 \\ \hline \\ \hline \end{array}$$

120.
$$\begin{array}{r} 7 \\ -\ 4 \\ \hline \\ \hline \end{array}$$

121.
$$\begin{array}{r} 3 \\ -\ 3 \\ \hline \\ \hline \end{array}$$

122.
$$\begin{array}{r} 6 \\ -\ 4 \\ \hline \\ \hline \end{array}$$

123.
$$\begin{array}{r} 6 \\ -\ 5 \\ \hline \\ \hline \end{array}$$

124.
$$\begin{array}{r} 4 \\ -\ 3 \\ \hline \\ \hline \end{array}$$

125.
$$\begin{array}{r} 17 \\ -\ 2 \\ \hline \\ \hline \end{array}$$

126.
$$\begin{array}{r} 12 \\ -\ 8 \\ \hline \\ \hline \end{array}$$

127.
$$\begin{array}{r} 8 \\ -\ 5 \\ \hline \\ \hline \end{array}$$

128.
$$\begin{array}{r} 14 \\ -\ 6 \\ \hline \\ \hline \end{array}$$

129.
$$\begin{array}{r} 12 \\ -\ 4 \\ \hline \\ \hline \end{array}$$

130.
$$\begin{array}{r} 11 \\ -\ 6 \\ \hline \\ \hline \end{array}$$

131.
$$\begin{array}{r} 1 \\ -\ 1 \\ \hline \\ \hline \end{array}$$

132.
$$\begin{array}{r} 14 \\ -\ 8 \\ \hline \\ \hline \end{array}$$

133.
$$\begin{array}{r} 12 \\ -\ 1 \\ \hline \\ \hline \end{array}$$

134.
$$\begin{array}{r} 16 \\ -\ 5 \\ \hline \\ \hline \end{array}$$

135.
$$\begin{array}{r} 15 \\ -\ 9 \\ \hline \\ \hline \end{array}$$

136.
$$\begin{array}{r} 13 \\ -\ 3 \\ \hline \\ \hline \end{array}$$

137.
$$\begin{array}{r} 13 \\ -\ 2 \\ \hline \\ \hline \end{array}$$

138.
$$\begin{array}{r} 12 \\ -\ 6 \\ \hline \\ \hline \end{array}$$

139.
$$\begin{array}{r} 18 \\ -\ 2 \\ \hline \\ \hline \end{array}$$

140.
$$\begin{array}{r} 13 \\ -\ 4 \\ \hline \\ \hline \end{array}$$

141.
$$\begin{array}{r} 6 \\ -\ 3 \\ \hline \\ \hline \end{array}$$

142.
$$\begin{array}{r} 16 \\ -\ 2 \\ \hline \\ \hline \end{array}$$

143.
$$\begin{array}{r} 10 \\ -\ 6 \\ \hline \\ \hline \end{array}$$

144.
$$\begin{array}{r} 12 \\ -\ 9 \\ \hline \\ \hline \end{array}$$

145.
$$\begin{array}{r} 9 \\ -\ 7 \\ \hline \\ \hline \end{array}$$

146.
$$\begin{array}{r} 9 \\ -\ 3 \\ \hline \\ \hline \end{array}$$

147.
$$\begin{array}{r} 5 \\ -\ 3 \\ \hline \\ \hline \end{array}$$

148.
$$\begin{array}{r} 15 \\ -\ 3 \\ \hline \\ \hline \end{array}$$

149.
$$\begin{array}{r} 17 \\ -\ 6 \\ \hline \\ \hline \end{array}$$

Advanced Addition

Find the sum.

150.
$$\begin{array}{r} 95 \\ +\ 10 \\ \hline \\ \hline \end{array}$$

151.
$$\begin{array}{r} 61 \\ +\ 21 \\ \hline \\ \hline \end{array}$$

152.
$$\begin{array}{r} 27 \\ +\ 30 \\ \hline \\ \hline \end{array}$$

153.
$$\begin{array}{r} 94 \\ +\ 73 \\ \hline \\ \hline \end{array}$$

154.
$$\begin{array}{r} 94 \\ +\ 17 \\ \hline \\ \hline \end{array}$$

155.
$$\begin{array}{r} 43 \\ +\ 49 \\ \hline \\ \hline \end{array}$$

156.
$$\begin{array}{r} 88 \\ +\ 67 \\ \hline \\ \hline \end{array}$$

157.
$$\begin{array}{r} 59 \\ +\ 39 \\ \hline \\ \hline \end{array}$$

158.
$$\begin{array}{r} 98 \\ +\ 60 \\ \hline \\ \hline \end{array}$$

159.
$$\begin{array}{r} 51 \\ +\ 37 \\ \hline \\ \hline \end{array}$$

160.
$$\begin{array}{r} 98 \\ +\ 29 \\ \hline \\ \hline \end{array}$$

161.
$$\begin{array}{r} 36 \\ +\ 57 \\ \hline \\ \hline \end{array}$$

162.
$$\begin{array}{r} 17 \\ +\ 93 \\ \hline \\ \hline \end{array}$$

163.
$$\begin{array}{r} 76 \\ +\ 48 \\ \hline \\ \hline \end{array}$$

164.
$$\begin{array}{r} 86 \\ +\ 29 \\ \hline \\ \hline \end{array}$$

165.
$$\begin{array}{r} 74 \\ +\ 75 \\ \hline \\ \hline \end{array}$$

166.
$$\begin{array}{r} 30 \\ +\ 15 \\ \hline \\ \hline \end{array}$$

167.
$$\begin{array}{r} 46 \\ +\ 37 \\ \hline \\ \hline \end{array}$$

168.
$$\begin{array}{r} 87 \\ +\ 77 \\ \hline \\ \hline \end{array}$$

169.
$$\begin{array}{r} 89 \\ +\ 41 \\ \hline \\ \hline \end{array}$$

170.
$$\begin{array}{r} 66 \\ +\ 83 \\ \hline \\ \hline \end{array}$$

171.
$$\begin{array}{r} 71 \\ +\ 61 \\ \hline \\ \hline \end{array}$$

172.
$$\begin{array}{r} 64 \\ +\ 11 \\ \hline \\ \hline \end{array}$$

173.
$$\begin{array}{r} 90 \\ +\ 83 \\ \hline \\ \hline \end{array}$$

174.
$$\begin{array}{r} 73 \\ +\ 90 \\ \hline \\ \hline \end{array}$$

175.
$$\begin{array}{r} 50 \\ +\ 35 \\ \hline \\ \hline \end{array}$$

176.
$$\begin{array}{r} 79 \\ +\ 66 \\ \hline \\ \hline \end{array}$$

177.
$$\begin{array}{r} 71 \\ + \ 77 \\ \hline \\ \hline \end{array}$$

178.
$$\begin{array}{r} 73 \\ + \ 13 \\ \hline \\ \hline \end{array}$$

179.
$$\begin{array}{r} 14 \\ + \ 33 \\ \hline \\ \hline \end{array}$$

180.
$$\begin{array}{r} 53 \\ + \ 56 \\ \hline \\ \hline \end{array}$$

181.
$$\begin{array}{r} 86 \\ + \ 75 \\ \hline \\ \hline \end{array}$$

182.
$$\begin{array}{r} 43 \\ + \ 67 \\ \hline \\ \hline \end{array}$$

183.
$$\begin{array}{r} 30 \\ + \ 83 \\ \hline \\ \hline \end{array}$$

184.
$$\begin{array}{r} 31 \\ + \ 32 \\ \hline \\ \hline \end{array}$$

185.
$$\begin{array}{r} 51 \\ + \ 31 \\ \hline \\ \hline \end{array}$$

186.
$$\begin{array}{r} 70 \\ + \ 21 \\ \hline \\ \hline \end{array}$$

187.
$$\begin{array}{r} 70 \\ + \ 26 \\ \hline \\ \hline \end{array}$$

188.
$$\begin{array}{r} 57 \\ + \ 74 \\ \hline \\ \hline \end{array}$$

189.
$$\begin{array}{r} 92 \\ + \ 17 \\ \hline \\ \hline \end{array}$$

190.
$$\begin{array}{r} 13 \\ + \ 92 \\ \hline \\ \hline \end{array}$$

191.
$$\begin{array}{r} 74 \\ + \ 95 \\ \hline \\ \hline \end{array}$$

Make Sum

What number should be added to the first number to make the second number?

192. $\begin{array}{r} 14 \\ + \\ \hline 32 \end{array}$

193. $\begin{array}{r} 24 \\ + \\ \hline 64 \end{array}$

194. $\begin{array}{r} 19 \\ + \\ \hline 69 \end{array}$

195. $\begin{array}{r} 21 \\ + \\ \hline 41 \end{array}$

196. $\begin{array}{r} 40 \\ + \\ \hline 52 \end{array}$

197. $\begin{array}{r} 30 \\ + \\ \hline 86 \end{array}$

198. $\begin{array}{r} 30 \\ + \\ \hline 91 \end{array}$

199. $\begin{array}{r} 50 \\ + \\ \hline 60 \end{array}$

200. $\begin{array}{r} 18 \\ + \\ \hline 24 \end{array}$

201. $\begin{array}{r} 12 \\ + \\ \hline 17 \end{array}$

202. $\begin{array}{r} 16 \\ + \\ \hline 17 \end{array}$

203. $\begin{array}{r} 43 \\ + \\ \hline 70 \end{array}$

204.
$$
\begin{array}{r}
24 \\
+ \\
\hline
25
\end{array}
$$

205.
$$
\begin{array}{r}
11 \\
+ \\
\hline
26
\end{array}
$$

206.
$$
\begin{array}{r}
70 \\
+ \\
\hline
87
\end{array}
$$

207.
$$
\begin{array}{r}
84 \\
+ \\
\hline
91
\end{array}
$$

208.
$$
\begin{array}{r}
45 \\
+ \\
\hline
91
\end{array}
$$

209.
$$
\begin{array}{r}
13 \\
+ \\
\hline
25
\end{array}
$$

210.
$$
\begin{array}{r}
12 \\
+ \\
\hline
15
\end{array}
$$

211.
$$
\begin{array}{r}
36 \\
+ \\
\hline
56
\end{array}
$$

212.
$$
\begin{array}{r}
85 \\
+ \\
\hline
93
\end{array}
$$

213.
$$
\begin{array}{r}
14 \\
+ \\
\hline
16
\end{array}
$$

214.
$$
\begin{array}{r}
55 \\
+ \\
\hline
62
\end{array}
$$

215.
$$
\begin{array}{r}
42 \\
+ \\
\hline
52
\end{array}
$$

216.
$$
\begin{array}{r}
12 \\
+ \\
\hline
14
\end{array}
$$

217.
$$
\begin{array}{r}
38 \\
+ \\
\hline
88
\end{array}
$$

218.
$$
\begin{array}{r}
43 \\
+ \\
\hline
74
\end{array}
$$

219.
$$\begin{array}{r} 47 \\ + \\ \hline 90 \end{array}$$

220.
$$\begin{array}{r} 14 \\ + \\ \hline 54 \end{array}$$

221.
$$\begin{array}{r} 16 \\ + \\ \hline 18 \end{array}$$

222.
$$\begin{array}{r} 37 \\ + \\ \hline 96 \end{array}$$

223.
$$\begin{array}{r} 39 \\ + \\ \hline 52 \end{array}$$

224.
$$\begin{array}{r} 76 \\ + \\ \hline 80 \end{array}$$

225.
$$\begin{array}{r} 25 \\ + \\ \hline 41 \end{array}$$

226.
$$\begin{array}{r} 14 \\ + \\ \hline 18 \end{array}$$

227.
$$\begin{array}{r} 72 \\ + \\ \hline 77 \end{array}$$

228.
$$\begin{array}{r} 13 \\ + \\ \hline 15 \end{array}$$

229.
$$\begin{array}{r} 14 \\ + \\ \hline 29 \end{array}$$

230.
$$\begin{array}{r} 74 \\ + \\ \hline 75 \end{array}$$

231.
$$\begin{array}{r} 36 \\ + \\ \hline 95 \end{array}$$

232.
$$\begin{array}{r} 27 \\ + \\ \hline 51 \end{array}$$

233.
$$\begin{array}{r} 56 \\ + \\ \hline 99 \end{array}$$

Advanced Addition

Find the sum.

234.
$$752 + 956$$

235.
$$567 + 493$$

236.
$$633 + 458$$

237.
$$928 + 132$$

238.
$$875 + 711$$

239.
$$468 + 421$$

240.
$$603 + 584$$

241.
$$751 + 136$$

242.
$$350 + 818$$

243.
$$547 + 179$$

244.
$$852 + 436$$

245.
$$798 + 475$$

246.
```
   895
 + 703
 _____

 _____
```

247.
```
   995
 + 534
 _____

 _____
```

248.
```
   937
 + 206
 _____

 _____
```

249.
```
   625
 + 887
 _____

 _____
```

250.
```
   533
 + 738
 _____

 _____
```

251.
```
   385
 + 334
 _____

 _____
```

252.
```
   986
 + 308
 _____

 _____
```

253.
```
   880
 + 505
 _____

 _____
```

254.
```
   832
 + 341
 _____

 _____
```

255.
```
   977
 + 968
 _____

 _____
```

256.
```
   777
 + 358
 _____

 _____
```

257.
```
   433
 + 439
 _____

 _____
```

258.
```
   977
 + 255
 _____

 _____
```

259.
```
   238
 + 348
 _____

 _____
```

260.
```
   257
 + 350
 _____

 _____
```

Advanced Subtraction

Find the difference.

261.
$$\begin{array}{r} 96 \\ - 29 \\ \hline \end{array}$$

262.
$$\begin{array}{r} 86 \\ - 56 \\ \hline \end{array}$$

263.
$$\begin{array}{r} 85 \\ - 52 \\ \hline \end{array}$$

264.
$$\begin{array}{r} 72 \\ - 67 \\ \hline \end{array}$$

265.
$$\begin{array}{r} 81 \\ - 68 \\ \hline \end{array}$$

266.
$$\begin{array}{r} 48 \\ - 27 \\ \hline \end{array}$$

267.
$$\begin{array}{r} 83 \\ - 64 \\ \hline \end{array}$$

268.
$$\begin{array}{r} 53 \\ - 30 \\ \hline \end{array}$$

269.
$$\begin{array}{r} 83 \\ - 39 \\ \hline \end{array}$$

270.
$$\begin{array}{r} 37 \\ - 28 \\ \hline \end{array}$$

271.
$$\begin{array}{r} 37 \\ - 32 \\ \hline \end{array}$$

272.
$$\begin{array}{r} 51 \\ - 48 \\ \hline \end{array}$$

273.
$$44 - 22$$

274.
$$74 - 41$$

275.
$$91 - 84$$

276.
$$51 - 47$$

277.
$$72 - 72$$

278.
$$82 - 51$$

279.
$$41 - 30$$

280.
$$41 - 37$$

281.
$$88 - 36$$

282.
$$98 - 68$$

283.
$$73 - 25$$

284.
$$66 - 46$$

285.
$$97 - 96$$

286.
$$82 - 49$$

287.
$$52 - 42$$

Multiple Addends

Find the sum.

288.
$$
\begin{array}{r}
922 \\
493 \\
+ \ 762 \\
\hline
\end{array}
$$

289.
$$
\begin{array}{r}
358 \\
899 \\
+ \ 423 \\
\hline
\end{array}
$$

290.
$$
\begin{array}{r}
852 \\
915 \\
+ \ 314 \\
\hline
\end{array}
$$

291.
$$
\begin{array}{r}
706 \\
487 \\
+ \ 15 \\
\hline
\end{array}
$$

292.
$$
\begin{array}{r}
115 \\
729 \\
+ \ 753 \\
\hline
\end{array}
$$

293.
$$
\begin{array}{r}
142 \\
623 \\
+ \ 27 \\
\hline
\end{array}
$$

294.
$$
\begin{array}{r}
639 \\
976 \\
+ \ 331 \\
\hline
\end{array}
$$

295.
$$
\begin{array}{r}
848 \\
490 \\
+ \ 12 \\
\hline
\end{array}
$$

296.
$$
\begin{array}{r}
78 \\
168 \\
+ \ 160 \\
\hline
\end{array}
$$

297.
```
   480
   906
+  495
_____
```

298.
```
   238
   375
+  167
_____
```

299.
```
   671
   735
+  255
_____
```

300.
```
   333
   236
+  526
_____
```

301.
```
   561
   293
+   77
_____
```

302.
```
   123
   422
+  453
_____
```

303.
```
   889
   519
+  374
_____
```

304.
```
    56
   359
+  154
_____
```

305.
```
   535
   185
+  516
_____
```

306.
```
   735
    84
+   13
_____
```

307.
```
   760
   290
+  741
_____
```

308.
```
   273
   382
+   19
_____
```

309.
$$818$$
$$479$$
$$+\ 123$$
_____

310.
$$846$$
$$443$$
$$+\ 992$$
_____

311.
$$508$$
$$807$$
$$+\ 397$$
_____

Basic Subtraction and Regrouping

Find the difference.

312.
$$580$$
$$-\ 142$$
_____

313.
$$890$$
$$-\ 636$$
_____

314.
$$410$$
$$-\ 390$$
_____

315.
$$540$$
$$-\ 349$$
_____

316.
$$920$$
$$-\ 175$$
_____

317.
$$200$$
$$-\ 108$$
_____

318.
$$210$$
$$-\ 100$$
_____

319.
$$450$$
$$-\ 372$$
_____

320.
$$810$$
$$-\ 782$$
_____

321.
$$\begin{array}{r} 560 \\ -\ 469 \\ \hline \\ \hline \end{array}$$

322.
$$\begin{array}{r} 500 \\ -\ 263 \\ \hline \\ \hline \end{array}$$

323.
$$\begin{array}{r} 890 \\ -\ 167 \\ \hline \\ \hline \end{array}$$

324.
$$\begin{array}{r} 660 \\ -\ 154 \\ \hline \\ \hline \end{array}$$

325.
$$\begin{array}{r} 760 \\ -\ 388 \\ \hline \\ \hline \end{array}$$

326.
$$\begin{array}{r} 900 \\ -\ 651 \\ \hline \\ \hline \end{array}$$

327.
$$\begin{array}{r} 680 \\ -\ 152 \\ \hline \\ \hline \end{array}$$

328.
$$\begin{array}{r} 390 \\ -\ 277 \\ \hline \\ \hline \end{array}$$

329.
$$\begin{array}{r} 640 \\ -\ 361 \\ \hline \\ \hline \end{array}$$

330.
$$\begin{array}{r} 280 \\ -\ 137 \\ \hline \\ \hline \end{array}$$

331.
$$\begin{array}{r} 440 \\ -\ 227 \\ \hline \\ \hline \end{array}$$

332.
$$\begin{array}{r} 860 \\ -\ 189 \\ \hline \\ \hline \end{array}$$

333.
$$\begin{array}{r} 900 \\ -\ 742 \\ \hline \\ \hline \end{array}$$

334.
$$\begin{array}{r} 990 \\ -\ 581 \\ \hline \\ \hline \end{array}$$

335.
$$\begin{array}{r} 270 \\ -\ 109 \\ \hline \\ \hline \end{array}$$

336.
$$370$$
$$-\ 187$$
_____

337.
$$990$$
$$-\ 747$$
_____

338.
$$470$$
$$-\ 251$$
_____

Multiple Operations

Find the solution.

339. $60 + 48 - 88 + 33 =$

340. $9 + 49 + 70 =$

341. $51 + 15 - 23 =$

342. $55 + 94 - 2 + 4 =$

343. $62 + 54 + 46 =$

344. $44 + 28 - 2 + 9 =$

345. $2 + 58 - 36 =$ ---

346. $85 + 67 - 82 =$ ---

347. $35 + 50 + 73 + 62 =$ -----------------------------------

348. $52 + 64 + 38 =$ ---

349. $66 + 86 + 5 + 66 =$ ----------------------------------

350. $80 + 93 - 26 + 8 =$ ----------------------------------

351. $6 + 95 - 64 + 36 =$ ----------------------------------

352. $6 + 6 + 91 + 44 =$ -----------------------------------

353. $4+ 41 + 95 + 91 =$ -------------------------------------

354. $80 + 59 + 89 + 52 =$ -----------------------------------

355. $47 + 41 - 67 =$ --------------------------------------

356. $35 +1- 41 =$ --

357. $60 +4- 30 =$ --

358. $64 + 38 + 76 + 49 =$ ---------------------------------

359. $80 + 64 - 34 =$ --------------------------------------

360. $75 +1- 96 + 62 =$ ------------------------------------

361. $71 + 49 + 62 =$ --

362. $77 + 44 + 7 =$ --

363. $70 + 22 + 56 =$ --

364. $72 + 57 + 7 + 96 =$ --

365. $53 + 19 + 39 =$ --

366. $27 + 90 + 8 =$ --

367. $64 + 36 - 61 + 7 =$ --

368. $42 + 68 - 15 =$ --

369. $94 + 24 + 76 + 61 =$ -------------------------------------

370. $17 + 79 - 62 =$ -------------------------------------

371. $27 + 80 + 97 =$ -------------------------------------

372. $79 + 89 - 91 =$ -------------------------------------

373. $93 + 71 - 39 =$ -------------------------------------

Multiple Operations: Missing Operators

Supply the missing operators to make the solution true.

374. $76 \underline{\quad} 27 \underline{\quad} 19 \underline{\quad} 79 = 201$

375. $49 \underline{\quad} 49 \underline{\quad} 19 = 117$

376. $62 \underline{\quad} 47 \underline{\quad} 11 = 120$

377. $42 \underline{\quad} 22 \underline{\quad} 47 \underline{\quad} 84 = 195$

378. $78 \underline{\quad} 78 \underline{\quad} 91 = 247$

379. $76 \underline{\quad} 53 \underline{\quad} 38 = 167$

380. $17 \underline{\quad} 41 \underline{\quad} 69 \underline{\quad} 27 = 16$

381. $53 \underline{\quad} 71 \underline{\quad} 37 \underline{\quad} 26 = 187$

382. $57 \underline{\quad} 68 \underline{\quad} 81 = 44$

383. $38 \underline{\quad} 68 \underline{\quad} 76 = 182$

384. $42 \underline{\quad} 32 \underline{\quad} 18 \underline{\quad} 30 = 122$

385. $43 \underline{\quad} 29 \underline{\quad} 39 \underline{\quad} 32 = 65$

386. $55 \underline{\quad} 84 \underline{\quad} 94 \underline{\quad} 74 = 119$

387. $1 \underline{\quad} 11 \underline{\quad} 31 \underline{\quad} 38 = 81$

388. $77 \underline{\quad} 60 \underline{\quad} 43 = 180$

389. $13 \underline{\quad} 72 \underline{\quad} 35 \underline{\quad} 58 = 178$

390. $92 \underline{\quad} 26 \underline{\quad} 77 \underline{\quad} 12 = 207$

391. $92 \underline{\quad} 6 \underline{\quad} 79 \underline{\quad} 30 = 49$

392. 5 ____ 53 ____ 71 ____ 42 = 171

393. 46 ____ 60 ____ 61 = 167

394. 33 ____ 25 ____ 99 ____ 84 = 241

395. 85 ____ 85 ____ 33 = 203

396. 79 ____ 70 ____ 26 ____ 95 = 218

397. 20 ____ 90 ____ 21 = 131

398. 94 ____ 56 ____ 4 ____ 7 = 153

399. 74 ____ 26 ____ 5 ____ 66 = 171

400. $17 \_\_\_ 46 \_\_\_ 26 = 89$

401. $76 \_\_\_ 54 \_\_\_ 89 = 41$

402. $21 \_\_\_ 30 \_\_\_ 68 = -17$

403. $84 \_\_\_ 77 \_\_\_ 17 \_\_\_ 28 = 172$

404. $19 \_\_\_ 13 \_\_\_ 87 = -55$

405. $50 \_\_\_ 73 \_\_\_ 59 = 182$

406. $46 \_\_\_ 7 \_\_\_ 85 = -32$

407. $97 \_\_\_ 86 \_\_\_ 88 \_\_\_ 10 = 281$

408. 11 _____ 97 _____ 68 = 40

409. 36 _____ 72 _____ 34 _____ 57 = 199

410. 67 _____ 8 _____ 54 = 21

411. 59 _____ 9 _____ 31 = 99

412. 43 _____ 59 _____ 41 = 143

413. 58 _____ 12 _____ 42 _____ 74 = 102

414. 92 _____ 75 _____ 43 _____ 16 = 226

415. 57 _____ 1 _____ 66 = -8

416. 59 ____ 59 ____ 32 ____ $48 = 134$

417. 40 ____ 88 ____ $45 = 83$

Word Problems - Addition

Solve.

418. Some apples were in the basket. Three more apples were added to the basket. Now there are nine apples. How many apples were in the basket before more apples were added?

_____

419. Sandra has nine more oranges than Janet. Janet has two oranges. How many oranges does Sandra have?

_____

420. Six red Books and six green Books are in the basket. How many Books are in the basket?

_____

421. Steven has nine plums and Paul has eight plums. How many plums do Steven and Paul have together?

_____

422. Two mangoes are in thc basket. Two more mangoes are put in the basket. How many mangoes are in the basket now?

_____

423. 12 pears were in the basket. Four are red and the rest are green. How many pears are green?

_____

424. Two peaches were in the basket. More peaches were added to the basket. Now there are nine peaches. How many peaches were added to the basket?

_____

425. Six Pencils were in the basket. More Pencils were added to the basket. Now there are 15 Pencils. How many Pencils were added to the basket?

_____

426. 11 bananas were in the basket. Five are red and the rest are green. How many bananas are green?

_____

427. Ellen has three more plums than Sandra. Sandra has three plums. How many plums does Ellen have?

_____

428. Adam has two bananas and Donald has two bananas. How many bananas do Adam and Donald have together?

_____

429. Five red Books and seven green Books are in the basket. How many Books are in the basket?

_____

430. Nine Pencils are in the basket. Five more Pencils are put in the basket. How many Pencils are in the basket now?

_____

431. Some oranges were in the basket. Seven more oranges were added to the basket. Now there are 12 oranges. How many oranges were in the basket before more oranges were added?

_____

432. 14 apples were in the basket. Five are red and the rest are green. How many apples are green?

_____

433. Jake has six peaches and Paul has five peaches. How many peaches do Jake and Paul have together?

_____

434. Jackie has seven more pears than Andria. Andria has four pears. How many pears does Jackie have?

_____

435. Nine red mangoes and nine green mangoes are in the basket. How many mangoes are in the basket?

_____

436. Some oranges were in the basket. Nine more oranges were added to the basket. Now there are 16 oranges. How many oranges were in the basket before more oranges were added?

_____

437. Six apples were in the basket. More apples were added to the basket. Now there are nine apples. How many apples were added to the basket?

_____

438. Seven bananas are in the basket. Seven more bananas are put in the basket. How many bananas are in the basket now?

_____

439. Nine Books were in the basket. Three are red and the rest are green. How many Books are green?

_____

440. Five Pencils were in the basket. More Pencils were added to the basket. Now there are 12 Pencils. How many Pencils were added to the basket?

_____

441. Nine red pears and six green pears are in the basket.
How many pears are in the basket?

_____

442. Robin has seven mangoes and Adam has nine mangoes.
How many mangoes do Robin and Adam have together?

_____

443. Some plums were in the basket. Eight more plums were
added to the basket. Now there are 16 plums. How many
plums were in the basket before more plums were
added?

_____

444. Nine peaches are in the basket. Three more peaches are
put in the basket. How many peaches are in the basket
now?

_____

445. Marin has two more Pencils than Ellen. Ellen has four
Pencils. How many Pencils does Marin have?

_____

446. Four apples were in the basket. Two are red and the rest are green. How many apples are green?

447. Some bananas were in the basket. Nine more bananas were added to the basket. Now there are 15 bananas. How many bananas were in the basket before more bananas were added?

Word Problems - Subtraction

Solve.

448. Four plums are in the basket. Three plums are taken out of the basket. How many plums are in the basket now?

449. Eight mangoes are in the basket. Eight are red and the rest are green. How many mangoes are green?

450. Adam has nine pears. David has nine pears. How many more pears does David have than Adam?

_____

451. Michele has one fewer peach than Jackie. Jackie has five peaches. How many peaches does Michele have?

_____

452. Some oranges were in the basket. Three oranges were taken from the basket. Now there are four oranges. How many oranges were in the basket before some of the oranges were taken?

_____

453. Eight bananas were in the basket. Some of the bananas were removed from the basket. Now there are four bananas. How many bananas were removed from the basket?

_____

454. Two Pencils are in the basket. Two are red and the rest are green. How many Pencils are green?

_____

455. Nine apples are in the basket. Six apples are taken out of the basket. How many apples are in the basket now?

456. Robin has two Books. Steven has seven Books. How many more Books does Steven have than Robin?

457. Andria has one fewer orange than Audrey. Audrey has six oranges. How many oranges does Andria have?

458. Some Pencils were in the basket. Eight Pencils were taken from the basket. Now there is one Pencil. How many Pencils were in the basket before some of the Pencils were taken?

459. Two peaches were in the basket. Some of the peaches were removed from the basket. Now there are zero peaches. How many peaches were removed from the basket?

460. Three mangoes are in the basket. Three mangoes are taken out of the basket. How many mangoes are in the basket now?

_____

461. David has four pears. Donald has five pears. How many more pears does Donald have than David?

_____

462. Seven apples are in the basket. Three are red and the rest are green. How many apples are green?

_____

463. Four bananas were in the basket. Some of the bananas were removed from the basket. Now there are two bananas. How many bananas were removed from the basket?

_____

464. Some plums were in the basket. Seven plums were taken from the basket. Now there is one plum. How many plums were in the basket before some of the plums were taken?

_____

465. Andria has one fewer Book than Jackie. Jackie has eight Books. How many Books does Andria have?

_____

466. Nine Books are in the basket. Nine Books are taken out of the basket. How many Books are in the basket now?

_____

467. Eight plums are in the basket. Six are red and the rest are green. How many plums are green?

_____

468. Some apples were in the basket. Eight apples were taken from the basket. Now there are zero apples. How many apples were in the basket before some of the apples were taken?

_____

469. Paul has four Pencils. Donald has four Pencils. How many more Pencils does Donald have than Paul?

_____

470. Jackie has zero fewer pears than Sandra. Sandra has three pears. How many pears does Jackie have?

_____

471. Two bananas were in the basket. Some of the bananas were removed from the basket. Now there are zero bananas. How many bananas were removed from the basket?

_____

472. Allan has three peaches. David has five peaches. How many more peaches does David have than Allan?

_____

473. Two oranges are in the basket. Two are red and the rest are green. How many oranges are green?

_____

474. Nine mangoes were in the basket. Some of the mangoes were removed from the basket. Now there are two mangoes. How many mangoes were removed from the basket?

_____

475. Audrey has seven fewer pears than Michele. Michele has nine pears. How many pears does Audrey have?

476. Seven Books are in the basket. Two Books are taken out of the basket. How many Books are in the basket now?

477. Some bananas were in the basket. Two bananas were taken from the basket. Now there are two bananas. How many bananas were in the basket before some of the bananas were taken?

Addition Across-Downs

Solve.

478.

3	+	7	+	0	=	
+		+		+		+
1	+	0	+	6	=	
+		+		+		+
4	+	5	+	8	=	
=		=		=		=
	+		+		=	

479.

4	+	9	+	7	=	
+		+		+		+
1	+	4	+	0	=	
+		+		+		+
3	+	9	+	5	=	
=		=		=		=
	+		+		=	

480.

5	+	4	+	1	=	
+		+		+		+
1	+	6	+	7	=	
+		+		+		+
8	+	9	+	6	=	
=		=		=		=
	+		+		=	

481.

1	+	2	+	5	=	
+		+		+		+
0	+	5	+	7	=	
+		+		+		+
8	+	1	+	6	=	
=		=		=		=
	+		+		=	

482.

5	+	0	+	5	=	
+		+		+		+
3	+	9	+	3	=	
+		+		+		+
1	+	0	+	8	=	
=		=		=		=
	+		+		=	

483.

5	+	8	+	6	=	
+		+		+		+
10	+	8	+	10	=	
+		+		+		+
9	+	9	+	8	=	
=		=		=		=
	+		+		=	

484.

3	+	2	+	0	=	
+		+		+		+
3	+	4	+	2	=	
+		+		+		+
10	+	3	+	2	=	
=		=		=		=
	+		+		=	

485.

2	+	3	+	4	=	
+		+		+		+
10	+	3	+	2	=	
+		+		+		+
4	+	8	+	9	=	
=		=		=		=
	+		+		=	

486.

0	+	1	+	6	=	
+		+		+		+
8	+	10	+	6	=	
+		+		+		+
10	+	3	+	0	=	
=		=		=		=
	+		+		=	

487.

6	+	1	+	6	=	
+		+		+		+
2	+	3	+	5	=	
+		+		+		+
1	+	9	+	7	=	
=		=		=		=
	+		+		=	

Subtraction Across-Downs

Solve.

488.

46	−	9	−	14	=	
−		−		−		−
16	−	0	−	9	=	
−		−		−		−
13	−	5	−	2	=	
=		=		=		=
	−		−		=	

489.

46	–	24	–	14	=	
–		–		–		–
22	–	10	–	5	=	
–		–		–		–
9	–	5	–	4	=	
=		=		=		=
	–		–		=	

490.

64	–	24	–	21	=	
–		–		–		–
24	–	10	–	8	=	
–		–		–		–
13	–	4	–	3	=	
=		=		=		=
	–		–		=	

491.

41	−	11	−	13	=	
−		−		−		−
19	−	3	−	7	=	
−		−		−		−
11	−	6	−	0	=	
=		=		=		=
	−		−		=	

492.

47	−	24	−	10	=	
−		−		−		−
9	−	7	−	0	=	
−		−		−		−
15	−	7	−	4	=	
=		=		=		=
	−		−		=	

493.

40	−	10	−	20	=	
−		−		−		−
8	−	0	−	7	=	
−		−		−		−
21	−	10	−	4	=	
=		=		=		=
	−		−		=	

494.

47	−	20	−	21	=	
−		−		−		−
20	−	10	−	8	=	
−		−		−		−
4	−	0	−	4	=	
=		=		=		=
	−		−		=	

495.

32	−	4	−	10	=	
−		−		−		−
19	−	1	−	8	=	
−		−		−		−
9	−	1	−	2	=	
=		=		=		=
	−		−		=	

496.

50	−	16	−	23	=	
−		−		−		−
16	−	7	−	7	=	
−		−		−		−
19	−	8	−	10	=	
=		=		=		=
	−		−		=	

497.

42	–	21	–	8	=	
–		–		–		–
8	–	6	–	1	=	
–		–		–		–
19	–	9	–	6	=	
=		=		=		=
	–		–		=	

Addition and Subtraction
Across-Downs

Solve.

498.

5	−	2	+	1	=	
−		+		−		+
2	+	5	−	1	=	
+		−		+		+
7	−	3	+	1	=	
=		=		=		=
	+		+		=	

499.

10	–	0	+	9	=	
–		+		–		+
0	+	3	–	3	=	
+		–		+		+
10	–	2	+	1	=	
=		=		=		=
	+		+		=	

500.

4	–	2	+	6	=	
–		+		–		+
2	+	7	–	3	=	
+		–		+		+
7	–	7	+	2	=	
=		=		=		=
	+		+		=	

501.

8	−	4	+	0	=	
−		+		−		+
4	+	1	−	0	=	
+		−		+		+
6	−	0	+	5	=	
=		=		=		=
	+		+		=	

502.

6	−	2	+	2	=	
−		+		−		+
2	+	0	−	0	=	
+		−		+		+
10	−	0	+	7	=	
=		=		=		=
	+		+		=	

503.

8	–	7	+	4	=	
–		+		–		+
7	+	0	–	0	=	
+		–		+		+
10	–	0	+	10	=	
=		=		=		=
	+		+		=	

504.

9	–	3	+	9	=	
–		+		–		+
3	+	7	–	4	=	
+		–		+		+
2	–	1	+	9	=	
=		=		=		=
	+		+		=	

505.

5	−	5	+	7	=	
−		+		−		+
5	+	0	−	0	=	
+		−		+		+
5	−	0	+	4	=	
=		=		=		=
	+		+		=	

506.

4	−	4	+	9	=	
−		+		−		+
4	+	7	−	1	=	
+		−		+		+
9	−	0	+	9	=	
=		=		=		=
	+		+		=	

507.

3	−	0	+	2	=	
−		+		−		+
0	+	4	−	2	=	
+		−		+		+
4	−	4	+	4	=	
=		=		=		=
	+		+		=	

Number Patterns

Find the pattern: Write two next numbers (and Pattern)

508. 84, 77, 70, 63, 56, 49, 42,   _____

509. 54, 52, 50, 48, 46, 44, 42,   _____

510. 24, 27, 30, 33, 36, 39, 42,   _____

511. 7, 11, 15, 19, 23, 27, 31,   —————————————

512. 50, 52, 54, 56, 58, 60, 62,   —————————————

513. 61, 64, 67, 70, 73, 76, 79,   —————————————

514. 19, 26, 33, 40, 47, 54, 61,   —————————————

515. 89, 81, 73, 65, 57, 49, 41,   —————————————

516. 96, 88, 80, 72, 64, 56, 48,   —————————————

517. 7, 16, 25, 34, 43, 52, 61,   —————————————

518. 65, 62, 59, 56, 53, 50, 47,   —————————————

519. 17, 24, 31, 38, 45, 52, 59,   ——————————

520. 25, 31, 37, 43, 49, 55, 61,   ——————————

521. 23, 31, 39, 47, 55, 63, 71,   ——————————

522. 6, 15, 24, 33, 42, 51, 60,   ——————————

523. 18, 26, 34, 42, 50, 58, 66,   ——————————

524. 76, 68, 60, 52, 44, 36, 28,   ——————————

525. 10, 13, 16, 19, 22, 25, 28,   ——————————

526. 3, 8, 13, 18, 23, 28, 33,   ——————————

527. 96, 89, 82, 75, 68, 61, 54,   _____

528. 22, 28, 34, 40, 46, 52, 58,   _____

529. 57, 53, 49, 45, 41, 37, 33,   _____

530. 65, 67, 69, 71, 73, 75, 77,   _____

531. 97, 91, 85, 79, 73, 67, 61,   _____

532. 62, 67, 72, 77, 82, 87, 92,   _____

533. 89, 82, 75, 68, 61, 54, 47,   _____

534. 31, 33, 35, 37, 39, 41, 43,   _____

535. 13, 20, 27, 34, 41, 48, 55,   _____

536. 16, 20, 24, 28, 32, 36, 40,   _____

537. 60, 55, 50, 45, 40, 35, 30,   _____

538. 76, 70, 64, 58, 52, 46, 40,   _____

539. 21, 30, 39, 48, 57, 66, 75,   _____

540. 1, 7, 13, 19, 25, 31, 37,   _____

541. 1, 8, 15, 22, 29, 36, 43,   _____

542. 7, 14, 21, 28, 35, 42, 49,   _____

543. 12, 15, 18, 21, 24, 27, 30,   _____

544. 78, 69, 60, 51, 42, 33, 24,   _____

545. 0, 4, 8, 12, 16, 20, 24,   _____

546. 59, 55, 51, 47, 43, 39, 35,   _____

547. 24, 29, 34, 39, 44, 49, 54,   _____

548. 84, 75, 66, 57, 48, 39, 30,   _____

549. 19, 25, 31, 37, 43, 49, 55,   _____

Secret Trails

Find the secret trail by adding the numbers.

550.

9	6	10
8	9	1
3	2	3

+ 21

551.

8	4	2
1	2	7
9	10	6

+ 27

552.

2	1	10
(7)	10	10
1	10	10

+ (37)

553.

1	10	8
10	6	8
(9)	1	6

+ (50)

554.

7	1	7
(2)	7	5
2	8	4

+ (18)

555.

(4)	5	2
8	8	8
6	3	9

+ (29)

556.

(10)	4	2
10	8	8
7	2	6

+ (30)

557.

(8)	9	5
10	8	2
5	8	1

+ (25)

558.

8	4	6
(8)	1	6
8	10	2

+ (35)

559.

2	6	2
(7)	6	3
9	8	4

+ (24)

560.

9	10	2
(10)	5	5
10	3	10

+ (42)

561.

(4)	8	2
5	1	7
9	5	9

+ (36)

562.

1	7	1
(9)	3	3
7	8	4

+ (32)

563.

4	7	5
(2)	6	9
2	6	3

+ (20)

564.

9	10	2
6	1	10
(8)	3	4

+ (41)

565.

9	9	10
1	8	4
(7)	6	4

+ (44)

566.

(10)	4	9
4	4	3
3	7	5

+ (26)

567.

(10)	2	8
5	5	7
3	2	6

+ (33)

568.

9	2	1
(8)	10	5
1	9	2

+ (28)

569.

8	2	3
(6)	7	8
4	7	6

+ (38)

570.

(9)	7	10
5	10	6
1	7	7

+ (39)

571.

(6)	10	9
8	2	9
7	4	3

+ (23)

572.

10	7	5
9	5	3
10	6	6

+ 31

573.

10	4	1
4	5	9
10	10	4

+ 22

574.

7	5	2
6	10	5
7	6	8

+ 43

575.

3	9	8
1	3	4
10	6	6

+ 14

576.

5	7	2
(10)	7	2
8	6	7

+ (40)

577.

4	7	1
10	7	2
(6)	9	4

+ (47)

578.

7	2	9
(1)	7	7
8	8	4

+ (19)

579.

(8)	4	2
2	3	1
9	6	1

+ (16)

580.

3	6	1
(5)	10	7
3	2	5

+ (34)

581.

(6)	4	10
8	4	8
5	4	8

+ (48)

582.

8	2	9
9	9	3
(9)	10	8

+ (45)

583.

7	9	3
10	5	2
(9)	5	6

+ (51)

584.

9	10	5
(6)	6	1
6	7	8
	+	(46)

585.

4	3	6
(3)	6	2
2	4	2
	+	(13)

586.

8	3	8
(8)	1	5
6	9	1
	+	(15)

587.

6	9	10
(4)	2	2
5	9	4
	+	(12)

588.

1	9	4
8	2	2
(4)	5	1
	+	(17)

589.

1	2	8
(2)	5	2
1	6	1
	+	(10)

Secret Trails

Find the secret trail by subtracting the numbers.

590.

6	6	5
3	8	10
25	2	7
	—	5

591.

1	10	4
36	2	9
9	7	10
	—	2

592.

9	2	5
(33)	8	2
2	5	9
	−	(7)

593.

2	4	3
1	8	7
(33)	1	9
	−	(8)

594.

1	10	9
3	7	6
(33)	9	10
	−	(1)

595.

(26)	5	3
1	8	4
2	8	5
	−	(9)

596.

6	1	7
(25)	5	2
8	5	5
	−	(3)

597.

2	9	8
1	6	5
(36)	10	5
	−	(10)

598.

3	2	5
1	4	2
(24)	9	1
	−	(4)

599.

2	9	8
(31)	6	9
9	10	6
	−	(6)

600.

37	9	4
9	1	2
10	1	8
	−	4

601.

1	1	8
34	6	3
2	3	6
	−	10

602.

27	8	6
3	9	4
9	10	7
	−	2

603.

1	1	4
6	9	5
36	4	9
	−	6

604.

27	7	3
1	10	7
8	3	4
	−	9

605.

7	3	6
17	2	4
3	1	6
	−	1

606.

41	4	4
9	10	8
9	10	2
	−	10

607.

10	7	8
24	9	5
3	7	7
	−	7

608.

2	4	10
(41)	5	9
9	8	5
	−	(5)

609.

9	5	7
7	6	4
(17)	2	3
	−	(2)

610.

9	5	3
3	2	3
(8)	1	4
	−	(3)

611.

(28)	1	4
9	3	10
1	8	4
	−	(2)

ANSWERS

Page 1:   Basic Addition

1. 8   2. 11   3. 6   4. 10   5. 3   6. 13   7. 5   8. 17   9. 9

10. 16   11. 15   12. 10   13. 9   14. 14   15. 14   16. 2   17. 14   18. 6

19. 3   20. 18   21. 8   22. 10   23. 12   24. 10   25. 11   26. 8   27. 7

28. 9   29. 6   30. 16   31. 8   32. 4   33. 11   34. 9   35. 9   36. 9

37. 7   38. 6   39. 8   40. 12   41. 11   42. 7

Page 4:   Make Sum

43. 1   44. 3   45. 2   46. 5   47. 0   48. 2   49. 0   50. 6   51. 2   52. 4   53. 4

54. 3   55. 2   56. 1   57. 4   58. 1   59. 1   60. 1   61. 1   62. 1   63. 2   64. 1

65. 6   66. 0   67. 7   68. 0   69. 3   70. 0   71. 5   72. 5   73. 3   74. 5   75. 0

76. 0   77. 2   78. 3   79. 0   80. 4   81. 6   82. 3   83. 4   84. 2

Page 7:   Pictorial Addition

85. 12   86. 8   87. 5   88. 6   89. 3   90. 10   91. 11   92. 17

93. 11   94. 10   95. 9   96. 6   97. 8   98. 8   99. 8   100. 10

101. 10   102. 10   103. 7   104. 7   105. 12   106. 5   107. 11

Page 9:   Basic Subtraction

108. 1   109. 14   110. 1   111. 9   112. 6   113. 0   114. 9   115. 2

116. 5   117. 17   118. 3   119. 6   120. 3   121. 0   122. 2   123. 1

124. 1   125. 15   126. 4   127. 3   128. 8   129. 8   130. 5   131. 0

132. 6   133. 11   134. 11   135. 6   136. 10   137. 11   138. 6   139. 16

140. 9   141. 3   142. 14   143. 4   144. 3   145. 2   146. 6   147. 2

148. 12   149. 11

Page 12:   Advanced Addition

150. 105   151. 82   152. 57   153. 167   154. 111   155. 92   156. 155

157. 98   158. 158   159. 88   160. 127   161. 93   162. 110   163. 124

164. 115   165. 149   166. 45   167. 83   168. 164   169. 130   170. 149

171. 132   172. 75   173. 173   174. 163   175. 85   176. 145   177. 148

178. 86   179. 47   180. 109   181. 161   182. 110   183. 113   184. 63

185. 82   186. 91   187. 96   188. 131   189. 109   190. 105   191. 169

Page 15:   Make Sum

192. 18   193. 40   194. 50   195. 20   196. 12   197. 56   198. 61   199. 10

200. 6   201. 5   202. 1   203. 27   204. 1   205. 15   206. 17   207. 7

208. 46   209. 12   210. 3   211. 20   212. 8   213. 2   214. 7   215. 10

216. 2   217. 50   218. 31   219. 43   220. 40   221. 2   222. 59   223. 13

224. 4   225. 16   226. 4   227. 5   228. 2   229. 15   230. 1   231. 59

232. 24   233. 43

Page 18:   Advanced Addition

234. 1,708   235. 1,060   236. 1,091   237. 1,060   238. 1,586   239. 889

240. 1,187   241. 887   242. 1,168   243. 726   244. 1,288   245. 1,273

246. 1,598   247. 1,529   248. 1,143   249. 1,512   250. 1,271   251. 719

252. 1,294   253. 1,385   254. 1,173   255. 1,945   256. 1,135   257. 872

258. 1,232   259. 586   260. 607

Page 20:   Advanced Subtraction

261. 67   262. 30   263. 33   264. 5   265. 13   266. 21   267. 19   268. 23

269. 44   270. 9   271. 5   272. 3   273. 22   274. 33   275. 7   276. 4

277. 0   278. 31   279. 11   280. 4   281. 52   282. 30   283. 48   284. 20

285. 1   286. 33   287. 10

Page 22:   Multiple Addends

288. 2,177   289. 1,680   290. 2,081   291. 1,208   292. 1,597   293. 792

294. 1,946   295. 1,350   296. 406   297. 1,881   298. 780   299. 1,661

300. 1,095   301. 931   302. 998   303. 1,782   304. 569   305. 1,236

306. 832   307. 1,791   308. 674   309. 1,420   310. 2,281   311. 1,712

Page 24:   Basic Subtraction and Regrouping

312. 438   313. 254   314. 20   315. 191   316. 745   317. 92   318. 110

319. 78   320. 28   321. 91   322. 237   323. 723   324. 506   325. 372

326. 249   327. 528   328. 113   329. 279   330. 143   331. 213   332. 671

333. 158   334. 409   335. 161   336. 183   337. 243   338. 219

Page 26:   Multiple Operations

339. 53   340. 128   341. 43   342. 151   343. 162   344. 79   345. 24

346. 70   347. 220   348. 154   349. 223   350. 155   351. 73   352. 147

353. 231   354. 280   355. 21   356. -5   357. 34   358. 227   359. 110

360. 42   361. 182   362. 128   363. 148   364. 232   365. 111   366. 125

367. 46   368. 95   369. 255   370. 34   371. 204   372. 77   373. 125

Page 30:   Multiple Operations: Missing Operators

374. $76 + 27 + 19 + 79 = 201$

375. $49 + 49 + 19 = 117$

376. $62 + 47 + 11 = 120$

377. $42 + 22 + 47 + 84 = 195$

378. $78 + 78 + 91 = 247$

379. $76 + 53 + 38 = 167$

380. $17 + 41 - 69 + 27 = 16$

381. $53 + 71 + 37 + 26 = 187$

382. $57 + 68 - 81 = 44$

383. $38 + 68 + 76 = 182$

384. $42 + 32 + 18 + 30 = 122$

385. $43 + 29 - 39 + 32 = 65$

386. $55 + 84 - 94 + 74 = 119$

387. $1 + 11 + 31 + 38 = 81$

388. $77 + 60 + 43 = 180$

389. $13 + 72 + 35 + 58 = 178$

390. $92 + 26 + 77 + 12 = 207$

391. $92 + 6 - 79 + 30 = 49$

392. $5 + 53 + 71 + 42 = 171$

393. $46 + 60 + 61 = 167$

394. $33 + 25 + 99 + 84 = 241$

395. $85 + 85 + 33 = 203$

396. $79 + 70 - 26 + 95 = 218$

397. $20 + 90 + 21 = 131$

398. $94 + 56 - 4 + 7 = 153$

399. $74 + 26 + 5 + 66 = 171$

400. $17 + 46 + 26 = 89$

401. $76 + 54 - 89 = 41$

402. $21 + 30 - 68 = -17$

403. $84 + 77 - 17 + 28 = 172$

404. $19 + 13 - 87 = -55$

405. $50 + 73 + 59 = 182$

406. $46 + 7 - 85 = -32$

407. $97 + 86 + 88 + 10 = 281$

408. $11 + 97 - 68 = 40$

409. $36 + 72 + 34 + 57 = 199$

410. $67 + 8 - 54 = 21$

411. $59 + 9 + 31 = 99$

412. $43 + 59 + 41 = 143$

413. $58 + 12 - 42 + 74 = 102$

414. $92 + 75 + 43 + 16 = 226$

415. $57 + 1 - 66 = -8$

416. $59 + 59 - 32 + 48 = 134$

417. $40 + 88 - 45 = 83$

Page 36:   Word Problems - Addition

418. 6   419. 11   420. 12   421. 17   422. 4   423. 8   424. 7   425. 9

426. 6   427. 6   428. 4   429. 12   430. 14   431. 5   432. 9   433. 11

434. 11   435. 18   436. 7   437. 3   438. 14   439. 6   440. 7   441. 15

442. 16   443. 8   444. 12   445. 6   446. 2   447. 6

Page 42:   Word Problems - Subtraction

448. 1   449. 0   450. 0   451. 4   452. 7   453. 4   454. 0   455. 3   456. 5   457. 5

458. 9   459. 2   460. 0   461. 1   462. 4   463. 2   464. 8   465. 7   466. 0   467. 2

468. 8   469. 0   470. 3   471. 2   472. 2   473. 0   474. 7   475. 2   476. 5   477. 4

Page 49:   Addition Across-Downs

478. 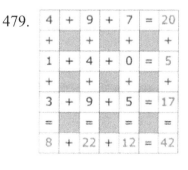   479. 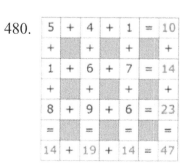   480.

481.
1	+	2	+	5	=	8
+		+		+		+
0	+	5	+	7	=	12
+		+		+		+
8	+	1	+	6	=	15
=		=		=		=
9	+	8	+	18	=	35

482.
5	+	0	+	5	=	10
+		+		+		+
3	+	9	+	3	=	15
+		+		+		+
1	+	0	+	8	=	9
=		=		=		=
9	+	9	+	16	=	34

483.
5	+	8	+	6	=	19
+		+		+		+
10	+	8	+	10	=	28
+		+		+		+
9	+	9	+	8	=	26
=		=		=		=
24	+	25	+	24	=	73

484.
3	+	2	+	0	=	5
+		+		+		+
3	+	4	+	2	=	9
+		+		+		+
10	+	3	+	2	=	15
=		=		=		=
16	+	9	+	4	=	29

485.
2	+	3	+	4	=	9
+		+		+		+
10	+	3	+	2	=	15
+		+		+		+
4	+	8	+	9	=	21
=		=		=		=
16	+	14	+	15	=	45

486.
0	+	1	+	6	=	7
+		+		+		+
8	+	10	+	6	=	24
+		+		+		+
10	+	3	+	0	=	13
=		=		=		=
18	+	14	+	12	=	44

487.
6	+	1	+	6	=	13
+		+		+		+
2	+	3	+	5	=	10
+		+		+		+
1	+	9	+	7	=	17
=		=		=		=
9	+	13	+	18	=	40

Page 55:   Subtraction Across-Downs

488.
46	−	9	−	14	=	23
−		−		−		−
16	−	0	−	9	=	7
−		−		−		−
13	−	5	−	2	=	6
=		=		=		=
17	−	4	−	3	=	10

489.
46	−	24	−	14	=	8
−		−		−		−
22	−	10	−	5	=	7
−		−		−		−
9	−	5	−	4	=	0
=		=		=		=
15	−	9	−	5	=	1

490.
64	−	24	−	21	=	19
−		−		−		−
24	−	10	−	8	=	6
−		−		−		−
13	−	4	−	3	=	6
=		=		=		=
27	−	10	−	10	=	7

491.
41	−	11	−	13	=	17
−		−		−		−
19	−	3	−	7	=	9
−		−		−		−
11	−	6	−	0	=	5
=		=		=		=
11	−	2	−	6	=	3

492.
47	−	24	−	10	=	13
−		−		−		−
9	−	7	−	0	=	2
−		−		−		−
15	−	7	−	4	=	4
=		=		=		=
23	−	10	−	6	=	7

493.
40	−	10	−	20	=	10
−		−		−		−
8	−	0	−	7	=	1
−		−		−		−
21	−	10	−	4	=	7
=		=		=		=
11	−	0	−	9	=	2

494.
47	−	20	−	21	=	6
−		−		−		−
20	−	10	−	8	=	2
−		−		−		−
4	−	0	−	4	=	0
=		=		=		=
23	−	10	−	9	=	4

495.
32	−	4	−	10	=	18
−		−		−		−
19	−	1	−	8	=	10
−		−		−		−
9	−	1	−	2	=	6
=		=		=		=
4	−	2	−	0	=	2

496.
50	−	16	−	23	=	11
−		−		−		−
16	−	7	−	7	=	2
−		−		−		−
19	−	8	−	10	=	1
=		=		=		=
15	−	1	−	6	=	8

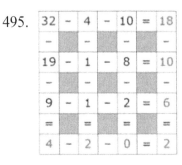

497.

42	−	21	−	8	=	13
−		−		−		−
8	−	6	−	1	=	1
−		−		−		−
19	−	9	−	6	=	4
=		=		=		=
15	−	6	−	1	=	8

Page 61:   Addition and Subtraction Across-Downs

498.

5	−	2	+	1	=	4
−		+		−		+
2	+	5	−	1	=	6
+		−		+		+
7	−	3	+	1	=	5
=		=		=		=
10	+	4	+	1	=	15

499.

10	−	0	+	9	=	19
−		+		−		+
0	+	3	−	3	=	0
+				+		+
10	−	2	+	1	=	9
=		=		=		=
20	+	1	+	7	=	28

500.

4	−	2	+	6	=	8
−		+		−		+
2	+	7	−	3	=	6
+		−		+		+
7	−	7	+	2	=	2
=		=		=		=
9	+	2	+	5	=	16

501.

8	−	4	+	0	=	4
−		+		−		+
4	+	1	−	0	=	5
+		−		+		+
6	−	0	+	5	=	11
=		=		=		=
10	+	5	+	5	=	20

502.

6	−	2	+	2	=	6
−		+		−		+
2	+	0	−	0	=	2
+		−		+		+
10	−	0	+	7	=	17
=		=		=		=
14	+	2	+	9	=	25

503.

8	−	7	+	4	=	5
−		+		−		+
7	+	0	−	0	=	7
+		−		+		+
10	−	0	+	10	=	20
=		=		=		=
11	+	7	+	14	=	32

504.

9	−	3	+	9	=	15
−		+		−		+
3	+	7	−	4	=	6
+		−		+		+
2	−	1	+	9	=	10
=		=		=		=
8	+	9	+	14	=	31

505.

5	−	5	+	7	=	7
−		+		−		+
5	+	0	−	0	=	5
+		−		+		+
5	−	0	+	4	=	9
=		=		=		=
5	+	5	+	11	=	21

506.

4	−	4	+	9	=	9
−		+		−		+
4	+	7	−	1	=	10
+		−		+		+
9	−	0	+	9	=	18
=		=		=		=
9	+	11	+	17	=	37

507.

3	−	0	+	2	=	5
−		+		−		+
0	+	4	−	2	=	2
+		−		+		+
4	−	4	+	4	=	4
=		=		=		=
7	+	0	+	4	=	11

Page 66:   Number Patterns

508. 35, 28 (- 7)   509. 40, 38 (- 2)   510. 45, 48 (+ 3)

511. 35, 39 (+ 4)   512. 64, 66 (+ 2)   513. 82, 85 (+ 3)

514. 68, 75 (+ 7)   515. 33, 25 (- 8)   516. 40, 32 (- 8)

517. 70, 79 (+ 9)   518. 44, 41 (- 3)   519. 66, 73 (+ 7)

520. 67, 73 (+ 6)   521. 79, 87 (+ 8)   522. 69, 78 (+ 9)

523. 74, 82 (+ 8)   524. 20, 12 (- 8)   525. 31, 34 (+ 3)

526. 38, 43 (+ 5)   527. 47, 40 (- 7)   528. 64, 70 (+ 6)

529. 29, 25 (- 4)   530. 79, 81 (+ 2)   531. 55, 49 (- 6)

532. 97, 102 (+ 5)   533. 40, 33 (- 7)   534. 45, 47 (+ 2)

535. 62, 69 (+ 7)   536. 44, 48 (+ 4)   537. 25, 20 (- 5)

538. 34, 28 (- 6)   539. 84, 93 (+ 9)   540. 43, 49 (+ 6)

541. 50, 57 (+ 7)   542. 56, 63 (+ 7)   543. 33, 36 (+ 3)

544. 15, 6 (- 9)   545. 28, 32 (+ 4)   546. 31, 27 (- 4)

547. 59, 64 (+ 5)   548. 21, 12 (- 9)   549. 61, 67 (+ 6)

Page 72:   Secret Trails

550.

9	6	10
8	9	1
3	2	3

+ 21

551.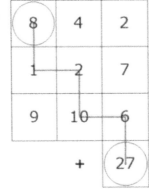

8	4	2
1	2	7
9	10	6

+ 27

552.

2	1	10
7	10	10
1	10	10

+ 37

553.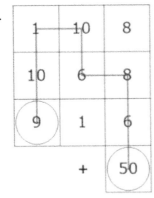

1	10	8
10	6	8
9	1	6

+ 50

554.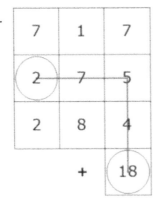

7	1	7
2	7	5
2	8	4

+ 18

555.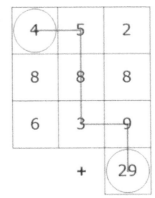

4	5	2
8	8	8
6	3	9

+ 29

556.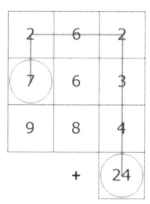

557.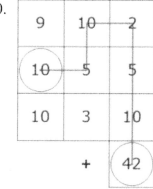

558.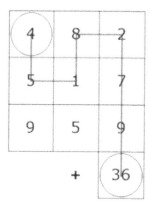

559.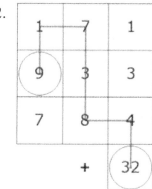

560.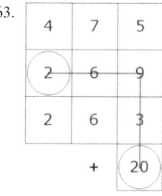

561.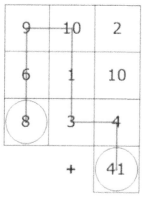

562.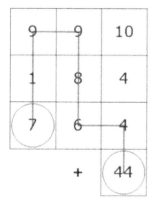

563.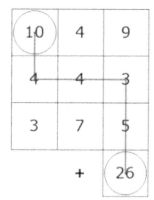

564.

565.

9	9	10
1	8	4
7	6	4

+ 44

566.

10	4	9
4	4	3
3	7	5

+ 26

567.

10	2	8
5	5	7
3	2	6

+ 33

568.
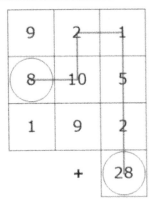

569.
8	2	3
6	7	8
4	7	6

+ 38

570.
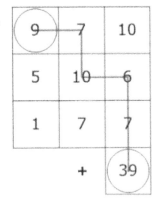

571.
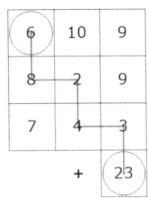

572.
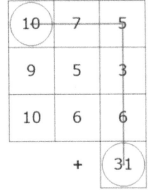

573.
10	4	1
4	5	9
10	10	4

+ 22

574.
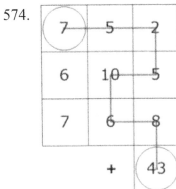

575.
3	9	8
1	3	4
10	6	6

+ 14

576.
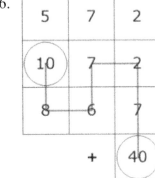

577.
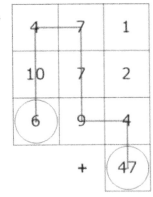

578.
7	2	9
1	7	7
8	8	4

+ 19

579.
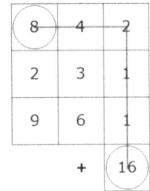

580.

3	6	1
5	10	7
3	2	5

+ 34

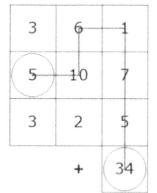

581.

6	4	10
8	4	8
5	4	8

+ 48

582.

8	2	9
9	9	3
9	10	8

+ 45

583.

7	9	3
10	5	2
9	5	6

+ 51

584.

9	10	5
6	6	1
6	7	8

+ 46

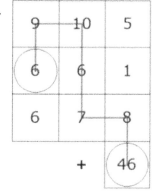

585.

4	3	6
3	6	2
2	4	2

+ 13

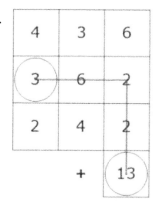

586.

8	3	8
8	1	5
6	9	1

+ 15

587.

6	9	10
4	2	2
5	9	4

+ 12

588.

1	9	4
8	2	2
4	5	1

+ 17

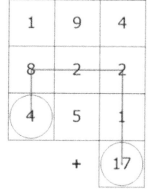

589.

1	2	8
2	5	2
1	6	1

+ 10

Page 83:   Secret Trails

590.

6	6	5
3	8	10
25	2	7

− 5

591.

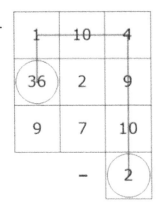

1	10	4
36	2	9
9	7	10

− 2

592.

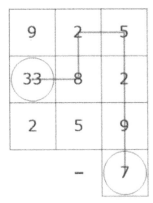

9	2	5
33	8	2
2	5	9

− 7

593.

2	4	3
1	8	7
33	1	9

− 8

594.

1	10	9
3	7	6
33	9	10

− 1

595.

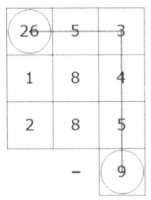

26	5	3
1	8	4
2	8	5

− 9

596.

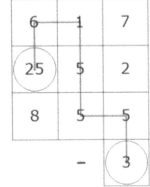

6	1	7
25	5	2
8	5	5

− 3

597.

2	9	8
1	6	5
36	10	5

− 10

598.

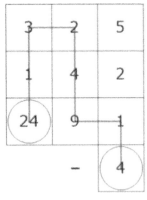

3	2	5
1	4	2
24	9	1

− 4

599.

2	9	8
31	6	9
9	10	6

− 6

600.

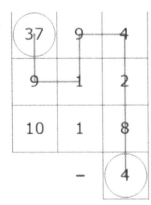

37	9	4
9	1	2
10	1	8

− 4

601.

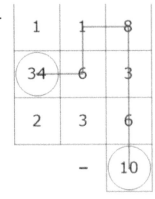

1	1	8
34	6	3
2	3	6

− 10

602.

27	8	6
3	9	4
9	10	7

− 2

603.

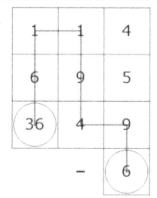

1	1	4
6	9	5
36	4	9

− 6

604.

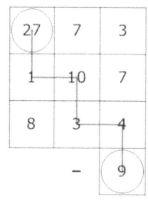

27	7	3
1	10	7
8	3	4

− 9

605.

7	3	6
17	2	4
3	1	6

− 1

606.

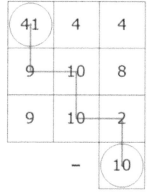

41	4	4
9	10	8
9	10	2

− 10

607.

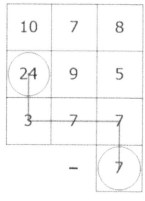

10	7	8
24	9	5
3	7	7

− 7

608.

2	4	10
41	5	9
9	8	5

− 5

609.

9	5	7
7	6	4
17	2	3

− 2

610.

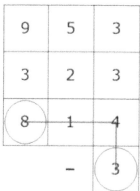

9	5	3
3	2	3
8	1	4

− 3

611.

28	1	4
9	3	10
1	8	4

− 2